MOLECULAR AND CELL BIOLOGY FOR COMPLETE BEGINNERS

The Definitive Guide on Exploring the Interactions Between Molecules and Cells and the Mechanisms of Cellular Function and Regulation

Mina Mong

Copyright@2024

TABLE OF CONTENTS

CHAPTER ONE 4
Introduction to Molecular and Cell Biology 4

CHAPTER TWO 10
The Building Blocks of Life: Atoms, Molecules, and Cells 10

CHAPTER THREE 16
The Structure and Function of Cells 16

CHAPTER FOUR 23
DNA, RNA, and Protein Synthesis 23

CHAPTER FIVE 30
Cellular Energy and Metabolism 30

CHAPTER SIX 37
Cellular Communication and Signalling 37

CHAPTER SEVEN 44
The Cell Cycle and Cell Division 44

CHAPTER EIGHT 50
Cellular Transport Mechanisms 50

CHAPTER NINE 57
Genetic Inheritance and Molecular Genetics 57

CHAPTER TEN 63
Introduction to Cell Differentiation and Development 63

CHAPTER ELEVEN 70

Introduction to Immunology: Cells and Molecules in the Immune Response 70

CHAPTER TWELVE 76

Advances in Molecular and Cell Biology 76

THE END 82

CHAPTER ONE

Introduction to Molecular and Cell Biology

The study of biological sciences is deeply rooted in the understanding of life at the molecular level and the cellular mechanisms that support it, offering profound insights into the complexities of living organisms. This introduction presents the definitions, importance, historical milestones, and applications of molecular and cell biology, highlighting its significant influence across various disciplines.

What is the study of biological molecules and their interactions?
Explanation and Significance
This field of science delves into the molecular foundations that underpin biological processes and activities. This research delves into the complex interactions among diverse cellular systems, focussing mainly on DNA, RNA, proteins, and other biomolecules. At its essence, the study of molecular biology aims to unravel the interactions between these molecules that propel cellular functions, such as replication, transcription, translation, and signalling within the cell.

The significance of this field of study is immense. This offers essential understanding of gene function, regulation, and the impact of molecular alterations on disease development. This discipline has played a crucial role in enhancing our comprehension of genetics, heredity, and the biochemical pathways that govern numerous physiological functions. This has led to significant advancements in our comprehension of genetic disorders, cancer mechanisms, and infectious diseases, fundamentally altering the landscape of medicine and treatment strategies.

A Concise Overview and Significant Findings

This discipline emerged as a distinct field in the mid-20th century, evolving from earlier areas such as biochemistry and genetics. Significant findings have influenced its evolution:

The discovery of the double helix structure of DNA in 1953 by James Watson and Francis Crick, informed by Rosalind Franklin's X-ray diffraction data, marked a significant milestone in the field. This finding established a foundational understanding of the mechanisms underlying the storage and replication of genetic information.

In 1958, Francis Crick presented a foundational concept that outlines the progression of genetic information, detailing how it transitions from DNA to RNA and ultimately to protein. This principle established a foundation for comprehending the intricacies of gene expression and its regulation.

The emergence of techniques for DNA manipulation in the 1970s, including the use of restriction enzymes and ligases, enabled researchers to construct recombinant DNA molecules. This advancement facilitated the cloning of genes and the creation of genetically modified organisms, transforming the fields of genetics and biotechnology.

Genome Sequencing (1990s-Present): The Human Genome Project, finalised in 2003, represented a significant endeavour to chart the complete human genome. This pursuit has greatly enhanced our comprehension of genetic mechanisms, inherited disorders, and the complexities of human biology.

These discoveries and others have established a foundational understanding, enabling deeper investigation into the intricate mechanisms of life at the molecular scale.

What constitutes the study of cells and their functions? Explanation and Importance

The exploration of cell biology, often referred to as cytology, involves examining the intricate structure, diverse functions, and dynamic behaviours of cells, which are the essential building blocks of life. This analysis explores the interactions between cells and their surroundings, the mechanisms of communication among them, and the execution of vital functions required for survival and reproduction.

The importance of studying cells is rooted in its capacity to clarify the mechanisms by which cells function both independently and in concert within tissues and living organisms. Grasping the intricacies of cellular mechanisms is essential for a range of applications, such as the advancement of pharmaceuticals, exploration of cancer, regenerative therapies, and insights into developmental biology.

Contextual Background and Significant Achievements

The study of cells boasts a profound history, marked by pivotal milestones that have greatly influenced our comprehension of cellular structures and functions:

Cell Theory (1839): Formulated by Theodor Schwann and Matthias Schleiden, this foundational principle asserts that all living organisms consist of cells, with the cell serving as the fundamental unit of life. This fundamental principle laid the groundwork for the exploration of biological sciences.

The evolution of microscopy from the 17th century to the present has enabled scientists to visualise cells for the very first time. Antonie van Leeuwenhoek's meticulous observations of unicellular organisms established a foundational framework for the fields of microbiology and cellular studies.

The identification of the nucleus in 1831 by Robert Brown represented a pivotal moment in the exploration of cellular structure and organisation.

Cellular Structures and Organelles (Late 19th to 20th Century): The discovery of distinct organelles, including mitochondria, Golgi apparatus, and lysosomes, has significantly enhanced our comprehension of cellular mechanisms.

The exploration of signalling pathways and mechanisms of cell communication has illuminated the ways in which cells synchronise their functions and react to external stimuli.

The historical milestones underscore the development of cell biology as an essential field within the life sciences, offering valuable insights into cellular functions and their implications for health and disease.

Why Explore the Intricacies of Life at the Cellular Level?
The exploration of molecular and cellular mechanisms presents a wealth of benefits and applications across diverse disciplines, establishing it as an essential domain of scientific research.

Utilisation in Healthcare
Disease Understanding and Treatment: The study of biological processes at the molecular level is crucial for comprehending the mechanisms underlying diseases. For example,

understanding the genetic foundations of cancer has resulted in targeted treatments that directly tackle the molecular alterations fuelling tumour development. Recent progress in diagnostic techniques allows for timely identification and tailored therapeutic strategies.

Gene Therapy: Advanced biological techniques, including CRISPR, show potential for rectifying genetic disorders by precisely targeting and modifying defective genes, providing possible solutions for ailments such as cystic fibrosis and sickle cell anaemia.

Biotechnology

Genetic Engineering: The principles of molecular biology serve as the essential basis for genetic engineering, facilitating the creation of genetically modified organisms (GMOs) that exhibit desirable characteristics, such as pest resistance in crops or enhanced nutritional value.

Biopharmaceuticals: The advent of recombinant DNA technology has transformed the landscape of biologic drug production, encompassing insulin, monoclonal antibodies, and vaccines, thereby enhancing therapeutic alternatives for a range of diseases.

Investigation and Innovation

Comprehending Essential Biological Mechanisms: Research in this field offers valuable insights into core processes like cell division, metabolism, and immune responses, enhancing our understanding of the essence of life.

Investigating Evolutionary Biology: Advanced techniques allow researchers to examine evolutionary connections and genetic variation across species, deepening our comprehension of biodiversity and conservation initiatives.

Innovative Advancements

Synthetic Biology: This interdisciplinary domain integrates concepts from biology, engineering, and computer science to innovate and assemble novel biological components and systems, offering promising applications in biofuels, environmental cleanup, and biomanufacturing.

CHAPTER TWO

The Building Blocks of Life: Atoms, Molecules, and Cells

At its core, life is constructed from the intricate interplay of atoms, molecules, and cells. The interactions among these components are intricately organised, resulting in the sophisticated structures and processes that characterise living organisms. Grasping the fundamental components of life—atoms, molecules, and cells—is essential for deciphering the complex processes that regulate biological activities. In this section, we will delve into the significance of atoms and molecules within biological systems, examine the primary categories of biological macromolecules, and provide an overview of cells as the essential units of life.

Fundamental Units and Compounds in Living Organisms
Atoms represent the fundamental building blocks of matter, preserving the distinct characteristics of each element. They function as the fundamental components of all matter, encompassing the molecules that constitute living organisms. An atom is composed of three fundamental subatomic particles:

Protons are positively charged particles found within the nucleus of an atom. The identity of an element is determined by its proton count; for instance, hydrogen possesses a single proton, while carbon contains six protons.
Neutrons are neutral particles located within the nucleus. Neutrons play a crucial role in determining the atomic mass of an atom, yet they do not influence its chemical properties.

Electrons are subatomic particles with a negative charge that reside in specific energy levels or shells surrounding the nucleus. Electrons serve as essential components in the processes of chemical reactions and the formation of bonds. In biological systems, atoms interact to create molecules via chemical bonds. The bonds that link atoms play a vital role in determining the structure and stability of biological molecules.

Types of Interactions in Biological Systems
Covalent bonds arise through the sharing of electron pairs between atoms. This bond is the most prevalent in biological systems and exhibits considerable strength, contributing to the stability of molecules such as DNA and proteins. In a water molecule (H_2O), oxygen engages in electron sharing with two hydrogen atoms, resulting in a stable molecular structure.

Ionic bonds arise when one atom transfers an electron to another atom, resulting in the formation of positively and negatively charged ions. The attraction between these oppositely charged ions leads to the formation of an ionic bond. Ionic bonds frequently occur in salts such as sodium chloride (NaCl), where sodium donates an electron to form Na^+, while chlorine accepts an electron to become Cl^-.

Hydrogen bonds, while weaker than covalent or ionic bonds, are crucial in the structure and function of biological molecules. A hydrogen bond occurs when a hydrogen atom, which is covalently bonded to an electronegative atom like oxygen or nitrogen, experiences an attraction to another electronegative atom. The significance of these bonds is paramount for the characteristics of water and the architecture of macromolecules such as DNA, where hydrogen bonds maintain the integrity of the double helix structure.

Large Biological Molecules within Cellular Structures
Life fundamentally consists of four essential categories of macromolecules: proteins, nucleic acids, lipids, and carbohydrates. These macromolecules play a crucial role in determining the architecture and operational capabilities of cells.

Proteins: Architecture and Role
Proteins are intricate, sizable structures made up of sequences of amino acids. These components play a crucial role in almost all biological processes, such as catalysing chemical reactions, offering structural support, transporting molecules, and enabling communication between cells.

Proteins consist of amino acids that are interconnected through peptide bonds, resulting in the formation of a polypeptide chain. The precise arrangement of amino acids within a protein dictates its conformation and biological role. Proteins assume unique three-dimensional conformations driven by the interactions among amino acids. These configurations can be classified into four distinct tiers:

Primary Structure: The sequential arrangement of amino acids in a polypeptide chain.
Secondary Structure: The local folding of polypeptide chains into configurations such as alpha helices and beta sheets.
The tertiary structure refers to the complete three-dimensional configuration of an individual polypeptide chain.
The quaternary structure refers to the organisation of several polypeptide chains into a functional protein complex.
Proteins carry out a diverse array of functions within cellular environments. Enzymes are proteins that speed up chemical reactions. Structural proteins, such as collagen, provide support to cells and tissues. Transport proteins, like hemoglobin, carry

molecules such as oxygen throughout the body. Proteins also play a role in immune defense (antibodies) and signal transduction (receptors).

Fundamentals of Nucleic Acids: An Overview of DNA and RNA
Nucleic acids serve as the fundamental molecules that store and convey genetic information. Nucleic acids can be classified into two primary categories: DNA, or deoxyribonucleic acid, and RNA, which stands for ribonucleic acid.

DNA serves as the fundamental genetic blueprint in nearly all forms of life. The structure comprises two strands of nucleotides intricately coiled into a double helix formation. A nucleotide consists of a phosphate group, a sugar known as deoxyribose, and one of four nitrogenous bases: adenine, thymine, guanine, or cytosine. The arrangement of these bases encodes genetic information. DNA plays a crucial role in transmitting genetic information across generations and orchestrating the production of proteins.

RNA plays a crucial role in the synthesis of proteins. Typically, it exists as a single strand and features ribose sugar in place of deoxyribose, with uracil substituting for thymine. RNA molecules, including messenger RNA (mRNA), transfer RNA (tRNA), and ribosomal RNA (rRNA), collaborate to convert the genetic code into functional proteins.

The Significance of Lipids and Carbohydrates in Cellular Architecture and Activity
Fatty substances: Lipids are non-polar molecules that encompass fats, oils, and phospholipids. These components are crucial for the formation of the cell membrane, serving as a protective barrier that separates the cell from its surrounding

environment. Phospholipids serve as the fundamental building blocks of the plasma membrane, creating a bilayer structure in which the hydrophobic tails orient inward, shielded from water, while the hydrophilic heads are positioned outward. Lipids play a crucial role in energy storage and provide insulation as well.

Carbohydrates consist of sugar molecules and function as a fundamental energy source for cellular processes. Monosaccharides such as glucose play a crucial role in cellular respiration, facilitating the production of ATP, which serves as the energy currency within the cell. Complex carbohydrates, including starch and glycogen, serve as energy reserves in plants and animals, respectively. Carbohydrates are integral to cell recognition and communication, exemplified by the presence of glycoproteins on the cell surface.

Cells: The Essential Building Blocks of Life
Cells serve as the fundamental unit of structure and function across all living organisms. All living entities, from solitary bacteria to complex humans, consist of cells that perform essential life functions. Cells can be classified into two main categories: prokaryotic and eukaryotic.

Prokaryotic and Eukaryotic Cells
Prokaryotic Cells: Prokaryotic cells represent fundamental, unicellular entities, including bacteria and archaea. These cells are devoid of a nucleus and membrane-bound organelles. Their genetic material resides in a region known as the nucleoid, and they possess a more straightforward internal architecture. Even with their basic structure, prokaryotes perform all vital life functions, such as metabolism, growth, and reproduction.

Eukaryotic cells exhibit a higher level of complexity and encompass a diverse range of organisms, including both unicellular and multicellular forms, such as animals, plants, fungi, and protists. Eukaryotic cells possess a distinct nucleus that contains their genetic material, along with various membrane-bound organelles that carry out specialised functions. The following organelles are included:

The nucleus serves as the repository for genetic information and orchestrates various cellular activities.
Mitochondria: Generate energy via the process of cellular respiration.
The endoplasmic reticulum and Golgi apparatus play crucial roles in the synthesis and transport of proteins.
Lysosomes are responsible for the degradation of waste materials and cellular debris.
The Concept of Cellular Theory and Its Importance
The concept of cell theory stands as a cornerstone of biological understanding, formulated in the 19th century by the pioneering scientists Matthias Schleiden, Theodor Schwann, and Rudolf Virchow. It outlines three primary principles:

Every living organism consists of one or more cells.
The cell serves as the fundamental unit of structure and organisation within living organisms.
Every cell originates from another cell that already exists.
This theory highlights the essential role of cells as the core components that constitute all living organisms. It established a foundation for comprehending the architecture and roles of living entities, spanning from the tiniest bacteria to the most massive mammals. The understanding that all biological processes take place within cells transformed the field of biology and established a foundation for investigating diseases, development, and evolution from a cellular perspective.

CHAPTER THREE

The Structure and Function of Cells

Cells represent the fundamental units of life, underpinning the architecture and operations of all living entities. Prokaryotic and eukaryotic cells possess distinct characteristics that allow them to perform vital biological functions. In this section, we will delve into the principles of cell theory, examine the distinctions between prokaryotic and eukaryotic cells, identify the essential components of eukaryotic cells, and investigate the processes involved in the storage, expression, and translation of genetic information into proteins.

Cell Theory
The concept of cell theory serves as a cornerstone in the field of biology, elucidating the characteristics of cells and their essential functions within living organisms. Formulated in the 19th century by researchers Matthias Schleiden, Theodor Schwann, and Rudolf Virchow, the theory of cells encompasses three fundamental principles:

Every living organism consists of one or more cells: Cells serve as the fundamental building blocks and operational units of all living entities, ranging from simple unicellular bacteria to intricate multicellular organisms such as humans.
The cell serves as the fundamental unit of life: Every biological process occurs within cells, establishing them as the smallest entity capable of executing all essential functions for life.
Every cell originates from an existing cell. Cells undergo division, a fundamental process that guarantees the perpetuation of life across generations.

The formulation and advancement of cell theory transformed the field of biology, enabling researchers to comprehend that all living entities possess a shared cellular basis.

Cells as the Fundamental Unit of Life

Cells represent the fundamental units of life, executing a diverse array of functions such as energy generation, reproduction, signalling, and the synthesis of vital molecules. Cells can be categorised into two primary types: prokaryotic and eukaryotic, each exhibiting unique structural and functional traits.

Comparison of Prokaryotic and Eukaryotic Cells

Prokaryotic Cells: Prokaryotic cells, which are present in organisms like bacteria and archaea, exhibit a simpler and smaller structure compared to eukaryotic cells. They do not possess a membrane-enclosed nucleus or any other membrane-enclosed organelles. Their genetic material is found in a region known as the nucleoid, and their cellular architecture is quite simple. Prokaryotes generally possess a sturdy cell wall that offers structural integrity and safeguards the organism.

Eukaryotic cells, present in organisms such as plants, animals, fungi, and protists, exhibit greater complexity and size compared to prokaryotic cells. Their structure includes a well-defined nucleus housing the genetic material of the cell, along with various membrane-bound organelles that carry out specialised functions. Eukaryotic cells can function independently as unicellular entities or contribute to the complexity of multicellular organisms. Their adaptability and intricate nature enable them to perform specialised functions, facilitating the development and sustenance of multicellular organisms.

Components of a Eukaryotic Cell

Eukaryotic cells are highly organized, with a variety of internal structures called organelles, each performing distinct functions essential for cellular operation.

Plasma Membrane: Architecture and Selective Permeability

The plasma membrane, often referred to as the cell membrane, serves as a delicate and adaptable barrier encasing the cell, meticulously controlling the passage of various substances into and out of the cellular environment. It is composed of a phospholipid bilayer with embedded proteins, cholesterol, and carbohydrates.

Selective Permeability: The plasma membrane is selectively permeable, meaning it allows certain molecules, such as water and gases, to pass freely while restricting the movement of larger or charged molecules. Transport proteins within the membrane facilitate the movement of essential molecules like glucose and ions, while other substances require active transport mechanisms.

Nucleus: DNA Storage and Gene Regulation

The nucleus serves as the central hub of the cell, housing the majority of its genetic material in the form of DNA. The structure is surrounded by a double membrane known as the nuclear envelope, which delineates it from the cytoplasm.

The DNA housed in the nucleus is meticulously organised into chromosomes, which carry the essential genetic instructions for the organism's development, function, and reproduction.

The nucleus is pivotal in the expression and regulation of genes, determining the activation or repression of specific genes in reaction to environmental stimuli or developmental signals. This regulation guarantees the precise production of proteins at the optimal time and in the correct amounts.

Cellular structures: Mitochondria, Endoplasmic Reticulum, Golgi Apparatus, Lysosomes, and others.

Mitochondria are often referred to as the powerhouse of the cell, playing a crucial role in energy production by generating adenosine triphosphate (ATP) through the process of cellular respiration. Mitochondria possess their own genetic material and play a crucial role in mechanisms such as apoptosis, which is the process of programmed cell death.

The Endoplasmic Reticulum (ER) is a complex network of membranes that plays a crucial role in the synthesis of proteins and lipids. There exist two distinct forms of endoplasmic reticulum:

The rough endoplasmic reticulum, characterised by its ribosome-studded appearance, plays a crucial role in the synthesis and folding of proteins.

Smooth ER: Characterised by the absence of ribosomes, it plays a crucial role in the synthesis of lipids and the detoxification of various substances.

The Golgi apparatus plays a crucial role in the modification, packaging, and distribution of proteins and lipids synthesised by the endoplasmic reticulum. It functions as a cellular distribution centre, guiding molecules to their ultimate locations.

Lysosomes are specialised organelles encased in membranes, housing a variety of digestive enzymes essential for cellular processes. They decompose waste products, repair damaged organelles, and eliminate foreign entities, contributing to the overall well-being of the cell.

Prokaryotic Cells

Prokaryotic cells represent a fundamental category of life, characterised by their simplicity as single-celled organisms that do not possess the membrane-bound organelles typical of eukaryotic cells.

Summary of Bacteria and Archaea

Bacteria represent the most prevalent and varied category of prokaryotic organisms, inhabiting an extensive array of environments. They can exhibit advantageous properties, such as those seen with gut microbiota, or detrimental effects that lead to disease manifestation.

Archaea are a unique group of prokaryotic organisms that frequently flourish in extreme habitats, including hot springs and salt lakes. These entities possess distinct biochemical characteristics that set them apart from bacteria.

Distinct Characteristics of Prokaryotic Cells

No Nucleus Present: In prokaryotic organisms, the genetic material resides in the nucleoid, which is an irregularly shaped area within the cell, instead of being contained within a nucleus. The structure of the cell wall in prokaryotes is characterised by a rigid composition, primarily consisting of peptidoglycan in bacteria, while archaea may utilise different materials. This architecture serves to provide both protection and structural integrity.

Genetic material, messenger molecules, and the foundational principles of molecular biology

The central dogma outlines the pathway of genetic information in a biological context: DNA → RNA → Protein. This process encompasses the replication of DNA, the transcription of DNA into RNA, and the translation of RNA into proteins.

DNA: The Foundation of Biological Existence

Configuration of DNA: The structure of DNA consists of two elongated strands that twist together to create a double helix configuration. Each strand is composed of nucleotides, which feature a phosphate group, a sugar (deoxyribose), and a nitrogenous base (adenine, thymine, cytosine, or guanine). The two strands are interconnected through hydrogen bonds formed between complementary bases, specifically adenine pairing with thymine and cytosine pairing with guanine.

DNA Duplication: During replication, the DNA double helix undergoes unwinding, with each strand acting as a template for the synthesis of a new complementary strand. This guarantees the precise replication and transmission of genetic material to progeny cells throughout the process of cell division.

RNA and Transcription Types of RNA:

mRNA (messenger RNA): Transmits genetic information from DNA to the ribosome, the site of protein synthesis.
tRNA (transfer RNA): Functions to transport amino acids to the ribosome throughout the process of translation.
rRNA (ribosomal RNA): Constitutes a component of the ribosome and facilitates the formation of peptide bonds in the process of protein synthesis.
Transcription refers to the mechanism through which the genetic information encoded in a DNA sequence is transcribed into an mRNA molecule. RNA polymerase attaches to the DNA and constructs the mRNA strand using the DNA template as a guide.

Protein Synthesis and Translation Processes
Genetic Code and Codons: The genetic code is composed of sequences of three nucleotides, known as codons, found in mRNA, with each codon designating a specific amino acid. For instance, the codon AUG is responsible for coding the amino acid methionine, which additionally functions as the initiation codon for the process of translation.

Ribosomes and Translation: The process of translation takes place at the ribosome, where messenger RNA (mRNA) is interpreted into a chain of amino acids. Transfer RNA molecules transport the appropriate amino acids to the ribosome, where the ribosome facilitates the creation of peptide bonds, resulting in a polypeptide chain that subsequently folds into a functional protein.

Alterations and Their Implications
Mutations represent alterations in the DNA sequence that can lead to diverse impacts on an organism's proteins and its overall functionality.

Categories of Genetic Alterations:
A point mutation involves the alteration of a single nucleotide, potentially affecting the resulting protein's structure or function.
Insertion: An additional nucleotide is incorporated, which may disrupt the reading frame of the genetic code and modify the resultant protein.
Deletion involves the removal of a nucleotide, potentially altering the reading frame and interfering with protein synthesis.
Consequences of Mutations: Mutations can result in dysfunctional proteins, potentially leading to conditions like cystic fibrosis or sickle cell anaemia. Nonetheless, it is important to recognise that not every mutation leads to detrimental effects; certain mutations may be neutral or advantageous, playing a role in genetic diversity and the process of evolution.

CHAPTER FOUR

DNA, RNA, and Protein Synthesis

The processes of DNA, RNA, and protein synthesis are fundamental to cellular life, illustrating the transfer of genetic information from DNA to RNA and ultimately to proteins. This section delves into the intricate processes that allow cells to duplicate their DNA, transcribe it into RNA, and translate RNA into functional proteins.

Structure and Replication of DNA
The intricate structure of DNA (deoxyribonucleic acid) serves as the foundational blueprint for all cellular functions. Comprehending the architecture of DNA and its replication processes is essential for elucidating the mechanisms of genetic inheritance and expression.

Double Helix Structure
The structure of DNA was initially elucidated in 1953 by James Watson and Francis Crick, with significant input from Rosalind Franklin. The structure of DNA is characterised by two elongated strands of nucleotides that are intricately coiled into a double helix formation. Every nucleotide consists of three essential components: a phosphate group, a sugar known as deoxyribose, and a nitrogenous base. DNA consists of four distinct nitrogenous bases:

Adenine (A)
Thymine (T)
Cytosine (C)
Guanine (G)

The two strands of DNA exhibit an antiparallel orientation, indicating that they extend in opposing directions. The bases on one strand engage in complementary pairing with those on the opposing strand: adenine aligns with thymine (A-T), while cytosine pairs with guanine (C-G). The hydrogen bonds that connect these base pairs consist of two bonds between A and T, and three bonds between C and G. The intricate nature of complementary base pairing allows DNA to effectively store genetic information and ensures accurate replication during cell division.

The process of DNA replication and the enzymes involved the process of DNA replication entails the meticulous copying of a cell's DNA prior to cell division, guaranteeing that each daughter cell inherits an exact replica of the genetic instructions. Replication adheres to a semi-conservative model, indicating that each newly formed DNA molecule comprises one original (parental) strand alongside one newly synthesized strand.

Processes involved in DNA Duplication:

Initiation: The process commences at designated sites on the DNA molecule known as origins of replication. A specific enzyme known as helicase facilitates the unwinding of the double helix by disrupting the hydrogen bonds that connect base pairs, thereby forming a replication fork.

Primer Synthesis: Following the unwinding of DNA, the enzyme primase catalyzes the synthesis of a short RNA primer that is complementary to the DNA template. This primer serves as an essential foundation for the synthesis of DNA.

Elongation involves the crucial enzyme DNA polymerase, which facilitates the addition of nucleotides to the expanding DNA strand. The enzyme responsible for DNA synthesis can only incorporate nucleotides in a 5' to 3' orientation, resulting in the continuous formation of the leading strand. On the lagging strand, DNA synthesis takes place in short segments known as Okazaki fragments, which are subsequently connected by DNA ligase.

Termination: Upon completion of the replication of the entire DNA molecule, the process comes to an end. The two newly formed DNA molecules consist of one strand from the parent and one strand that has been newly synthesised, thereby maintaining genetic continuity.

Essential enzymes that play a role in replication consist of:
Helicase: Responsible for unwinding the DNA double helix structure.
Primase: Responsible for the synthesis of RNA primers.
DNA polymerase: Incorporates nucleotides to synthesize the new DNA strand.
Ligase: Joins the discontinuities between Okazaki fragments on the lagging strand.
Transcription and RNA Modification
Transcription involves the transformation of genetic information encoded in DNA into a corresponding RNA molecule. This RNA serves as a template for protein synthesis during translation. The fundamental principle of biological information transfer is represented as follows: DNA → RNA → Protein.

From DNA to RNA: The Process of Transcription Procedure
Transcription occurs within the nucleus of eukaryotic cells, where a single-stranded RNA molecule is synthesized using the DNA template as a guide. The procedure involves three primary phases:

Initiation occurs when RNA polymerase attaches to a designated area of the DNA known as the promoter, indicating the commencement of a gene. Transcription factors, which are proteins, play a crucial role in aiding RNA polymerase to identify and attach to the promoter region.

During elongation, RNA polymerase traverses the DNA template, unwinding the double helix while synthesising a complementary RNA strand in the 5' to 3' direction. In RNA, the base uracil (U) takes the place of thymine (T), resulting in adenine pairing with uracil during the process of transcription.

Termination occurs when RNA polymerase encounters a termination sequence on the DNA, indicating the conclusion of the gene. The newly synthesised RNA strand, referred to as pre-mRNA in eukaryotes, is released.

Types and Functions of mRNA, tRNA, and rRNA
mRNA serves as the carrier of genetic instructions, transporting information from the DNA located in the nucleus to the ribosomes in the cytoplasm, the site of protein synthesis. It comprises codons, which are sequences of three nucleotides that correspond to particular amino acids or signals during the process of translation.

tRNA molecules serve as carriers for amino acids, delivering them to the ribosome in the process of protein synthesis. Every tRNA molecule possesses an anticodon that pairs specifically with a corresponding mRNA codon, along with a linked amino acid that it carries.

rRNA (Ribosomal RNA): rRNA serves as a fundamental structural element of ribosomes, the intricate molecular entities responsible for protein synthesis. Ribosomes consist of rRNA and proteins, playing a crucial role in ensuring the accurate pairing of tRNA with mRNA codons throughout the process of translation.

RNA Modification in Eukaryotic Cells

Prior to the translation of mRNA into a protein, it must first undergo a series of modifications within eukaryotic cells:

In the process of capping, a modified guanine nucleotide is attached to the 5' end of the mRNA molecule. This modification serves to safeguard the mRNA from degradation while also promoting efficient binding to ribosomes.

Polyadenylation involves the addition of a poly-A tail, which consists of a sequence of adenine nucleotides, to the 3' end of the mRNA. This modification serves to stabilise the mRNA and facilitates its export from the nucleus.

During splicing, the non-coding segments of mRNA, known as introns, are excised, while the coding sequences, referred to as exons, are ligated together. The processed mRNA is now prepared for the translation process.

Translation and Protein Synthesis

Translation refers to the mechanism through which the genetic information encoded in mRNA is utilised to produce a protein. This process takes place in the cytoplasm at the ribosomes, where mRNA is translated into a chain of amino acids.

The Blueprint of Life

The genetic code is composed of triplet sequences of nucleotides known as codons, where each codon is linked to a particular amino acid. There exists a total of 64 codons, among which 61 are responsible for coding amino acids, while 3 function as stop signals that conclude the process of translation. The genetic code exhibits universality, indicating its consistency across nearly all living organisms.

Ribosomes and the Mechanism of Translation
The process of translation can be categorised into three distinct stages:

Initiation occurs when the small subunit of the ribosome associates with the 5' end of the mRNA molecule. This is subsequently followed by the binding of the initiator tRNA, which is responsible for delivering methionine, the amino acid that marks the beginning of the protein synthesis process. The assembly of the large ribosomal subunit leads to the formation of a complete ribosome.

During elongation, the ribosome advances along the mRNA, pairing each codon with its complementary tRNA anticodon. The tRNA molecules transport particular amino acids to the ribosome, which facilitates the formation of peptide bonds among the amino acids, resulting in the elongation of a polypeptide chain.

Termination occurs when the ribosome comes across a stop codon (UAA, UAG, or UGA), signalling the conclusion of translation. The finished polypeptide chain is liberated, and the ribosome undergoes disassembly.

Post-Translational Modifications and Protein Folding
After a protein is synthesised, it frequently experiences additional modifications and folding to attain its active conformation.

Significance of Protein Architecture for Activity
Proteins are required to adopt precise three-dimensional conformations to perform their biological roles effectively. The

arrangement of amino acids within a protein is crucial, as it ultimately shapes the protein's final conformation, which subsequently governs its biological role. Misfolded proteins can result in a loss of function or contribute to disease, as observed in conditions such as Alzheimer's and cystic fibrosis.

Chaperone Proteins and Folding Mechanisms
Chaperone proteins play a crucial role in ensuring that newly synthesised proteins fold correctly. They assist in averting misfolding and aggregation by interacting with polypeptides and directing them along the appropriate folding pathways. Chaperones are essential for preserving cellular integrity by facilitating the proper folding and stability of proteins in their functional forms.

Beyond the process of folding, numerous proteins experience post-translational modifications, including phosphorylation, glycosylation, or the incorporation of lipid groups, all of which are essential for their functionality, positioning, and durability.

CHAPTER FIVE

Cellular Energy and Metabolism

Cells necessitate a continuous influx of energy to execute a range of functions, from synthesising intricate molecules to facilitating mechanical activities. Metabolism includes the entirety of chemical reactions occurring within cells that are essential for sustaining life. This section delves into the fundamental principles of metabolism, examining the mechanisms of energy production via cellular respiration and photosynthesis, as well as the contributions of organelles such as mitochondria and chloroplasts in the generation of energy.

Overview of Metabolic Processes
Metabolism encompasses the entirety of biochemical reactions taking place within a living organism. The reactions can be classified into two main categories: anabolism and catabolism.

Anabolic pathways are characterised by the synthesis of complex molecules from simpler precursors, necessitating an input of energy. For instance, the assembly of proteins from amino acids or the construction of nucleotides into DNA represents anabolic processes. Anabolism plays a crucial role in facilitating growth, repair, and maintenance processes within cells.

Catabolism: The processes involved in catabolic pathways facilitate the decomposition of intricate molecules into more elementary forms, liberating energy that can be harnessed by the cell. The degradation of glucose in the context of cellular respiration represents a catabolic pathway that produces ATP, which serves as the main energy currency for the cell.

The interplay between anabolic and catabolic pathways creates metabolic cycles that sustain the equilibrium of energy and substances within cellular environments.

ATP: The Energy Currency of the Cell
Adenosine triphosphate (ATP) serves as the main molecule for energy storage and transfer within cells. It drives nearly every cellular function, ranging from muscle contraction to the active transport of substances across membranes.

Architecture and Role of ATP
ATP consists of three fundamental components:

Adenine is classified as a nitrogenous base.
Ribose is a pentose sugar composed of five carbon atoms.
Three Phosphate Units: The connections among these phosphate groups represent high-energy interactions. The hydrolysis of ATP, involving the removal of a phosphate group, results in the release of a considerable amount of energy that the cell can harness for a range of functions.
The breakdown of ATP leads to the production of ADP (adenosine diphosphate) along with an inorganic phosphate (Pi). The energy generated during this process is harnessed for various cellular functions, including chemical reactions, molecular transport, and mechanical work. ATP is perpetually synthesised from ADP and inorganic phosphate during cellular respiration, ensuring a continuous energy supply for the cell.

ATP Production (Metabolic Processes)
Cellular respiration refers to the mechanism through which cells produce ATP by metabolising glucose and various other molecules. This process takes place in both the cytoplasm and

mitochondria of the cell, comprising three primary stages:

Glycolysis represents the initial phase of cellular respiration, occurring within the cytoplasm. The process of glycolysis involves the conversion of a single glucose molecule, which consists of six carbon atoms, into two pyruvate molecules, each containing three carbon atoms. This process yields a net gain of 2 ATP molecules, accompanied by 2 NADH, which serve as electron carriers for subsequent stages.

The Krebs Cycle, also known as the Citric Acid Cycle, follows glycolysis. Here, pyruvate is transported into the mitochondria, where it undergoes further degradation within the cycle. In this cycle, acetyl-CoA, which originates from pyruvate, undergoes oxidation, resulting in the production of carbon dioxide (CO_2), 2 ATP molecules, and high-energy electron carriers such as NADH and $FADH_2$.

The concluding phase of cellular respiration takes place within the inner membrane of the mitochondria and encompasses two primary processes: the electron transport chain (ETC) and chemiosmosis. NADH and $FADH_2$ transfer electrons to the electron transport chain, where they are sequentially conveyed through a series of proteins. The movement of electrons facilitates the translocation of protons (H^+ ions) across the mitochondrial membrane, resulting in the establishment of a proton gradient. During chemiosmosis, protons traverse back into the mitochondrial matrix via ATP synthase, facilitating the production of ATP. This phase generates the majority of ATP— up to 34 molecules for each glucose molecule.

Under optimal conditions, the complete process of cellular respiration can yield as many as 38 ATP molecules from a single glucose molecule.

Glycolysis, the Krebs Cycle, and oxidative phosphorylation
The three stages of cellular respiration—glycolysis, the Krebs cycle, and oxidative phosphorylation—are intricately linked, creating a seamless pathway for energy extraction from glucose.

Glycolysis takes place in the cytoplasmic environment.
Does not necessitate oxygen (anaerobic process).
Catabolises a single glucose molecule into two pyruvate molecules.
Results in a net yield of 2 ATP and 2 NADH.
The Krebs Cycle, also known as the Citric Acid Cycle, occurs within the mitochondrial matrix.
Depends on the presence of oxygen (aerobic process).
Oxidises acetyl-CoA, resulting in the production of CO_2, NADH, $FADH_2$, and a minor yield of ATP.
Essential intermediates are generated for various metabolic pathways.
Oxidative Phosphorylation: Takes place within the inner mitochondrial membrane.
The transfer of electrons from NADH and $FADH_2$ through the electron transport chain leads to the establishment of a proton gradient.
ATP is produced as protons traverse through ATP synthase.
The efficiency of these processes in ATP generation is crucial for powering cellular functions. In the absence of oxygen, cells resort to fermentation to generate ATP; however, this mechanism is significantly less efficient compared to aerobic respiration.

The Function of Mitochondria in Generating Energy

Mitochondria are recognised as the energy-producing centres of the cell, serving as the location for aerobic respiration and the synthesis of ATP. The inner mitochondrial membrane houses the electron transport chain and ATP synthase, both of which are crucial for the process of oxidative phosphorylation.

The inner mitochondrial membrane exhibits extensive folding into cristae, enhancing the surface area for the electron transport chain's activity. The structure contains the protein assemblies that play a crucial role in electron transfer and ATP synthesis.

The mitochondrial matrix serves as the crucial environment for the Krebs cycle, facilitating the oxidation of acetyl-CoA and the production of high-energy electron carriers, namely NADH and $FADH_2$.

Mitochondria are crucial in regulating cell death (apoptosis) and calcium storage, highlighting their significance beyond mere ATP production.

Photosynthesis (For Plants and Organisms Capable of Photosynthesis)

Photosynthesis represents a fundamental biochemical process wherein plants, algae, and certain bacteria harness light energy, transforming it into chemical energy that is subsequently stored as glucose. This process is fundamental for sustaining life on our planet, serving as the primary energy source for the majority of ecosystems.

Reactions Dependent on Light and Reactions Independent of Light

Photosynthesis takes place in two distinct phases:

The light-dependent reactions occur within the thylakoid membranes of chloroplasts. Light is essential, and the process includes these steps:

Photon Absorption: The green pigment in plants, chlorophyll, captures light energy emitted by the sun. This energy stimulates the electrons within the chlorophyll molecule.
Photolysis involves harnessing light energy to cleave water molecules, resulting in the production of oxygen, protons, and electrons. Oxygen is produced and released as a byproduct.
The electron transport chain facilitates the transfer of excited electrons, leading to the production of ATP and NADPH via chemiosmosis, akin to mitochondrial processes.
The reactions that take place in the stroma of the chloroplasts are not dependent on light. The Calvin cycle utilises ATP and NADPH generated in the light-dependent reactions to transform CO_2 into glucose via a sequence of enzyme-mediated processes. The essential stages encompass:

Carbon Fixation: The enzyme RuBisCO catalyses the conversion of CO_2 into a stable organic molecule.
Reduction: The energy derived from ATP and NADPH facilitates the transformation of fixed carbon into glucose.
The cycle facilitates the regeneration of RuBP (ribulose bisphosphate), enabling the continuation of the process.

Chloroplasts and Their Function in Energy Generation
Chloroplasts serve as the organelles that facilitate the process of photosynthesis in plant cells and algae. These entities possess distinct structures that facilitate the absorption of light energy, transforming it into chemical energy.

Thylakoids are specialised membrane-enclosed structures that facilitate the light-dependent reactions of photosynthesis. Thylakoids are organised into structures known as grana.
Stroma: The aqueous environment encasing the thylakoids,

serving as the site for the Calvin cycle (light-independent reactions).

Chloroplasts house pigments like chlorophyll, which are adept at absorbing light primarily in the blue and red regions of the spectrum. The energy harnessed during photosynthesis is stored in the form of glucose, which can subsequently be utilised in cellular respiration to produce ATP.

CHAPTER SIX

Cellular Communication and Signalling

Cell communication and signalling are essential mechanisms that allow cells to synchronise their activities, react to environmental changes, and uphold homeostasis. This section will delve into the significance of cellular communication, examining various signalling mechanisms, the function of receptors, and the intricacies of signal transduction pathways, along with specific instances of cell signalling pathways.

The Importance of Cellular Communication
Cellular communication is crucial for sustaining homeostasis, the stable internal environment necessary for optimal functioning. Effective communication enables cells to synchronise their activities, adapt to environmental changes, and manage physiological processes. Here are several important reasons why communication within cells is essential:

Collaboration of Activities: Cells within multicellular entities are required to cooperate in order to execute intricate functions. For instance, muscle cells require communication with nerve cells to synchronise movement, whereas immune cells need to signal one another to initiate a defence against pathogens.

Cells must react to a range of external signals, including hormones and nutrients, to adjust to fluctuations in their surroundings. This is crucial for mechanisms such as growth, differentiation, and responses to stress.

The regulation of homeostasis is fundamentally influenced by cellular communication, which is essential for maintaining balance within biological systems. The endocrine system, for instance, secretes hormones that govern metabolism, growth, and various essential functions. Cells utilize feedback mechanisms to fine-tune their activities, ensuring that they operate under optimal conditions.

During the process of development, cells engage in communication to orchestrate differentiation and the formation of tissues. In the event of injury, intricate signaling pathways are activated to initiate repair mechanisms and orchestrate the healing process.

Categories of Cellular Communication
Cells utilize a range of signaling mechanisms for communication, which can be categorized into four primary types:

In paracrine signaling, cells secrete signaling molecules that influence adjacent target cells. This form of signaling typically occurs locally and encompasses growth factors, cytokines, and neurotransmitters. For instance, immune cells utilize paracrine signaling to convey information and stimulate adjacent immune cells during an immune response.

Endocrine Signaling: In this process, hormones are released into the bloodstream by endocrine glands, allowing them to traverse significant distances to interact with target cells across the body. This signaling mechanism operates at a slower pace, yet it produces enduring effects. An illustrative case is insulin, secreted by the pancreas, which plays a crucial role in the regulation of glucose concentrations in the bloodstream.

In autocrine signalling, cells secrete signalling molecules that interact with receptors on their own membranes, thereby modulating their own activities. This phenomenon is prevalent in immune cells, as they possess the ability to self-activate in reaction to infection.

Direct Contact: Cells engage in communication through direct contact, employing specialized proteins known as cell adhesion molecules (CAMs). This form of signaling plays a vital role in processes such as embryonic development, where cells need to identify and engage with one another to create tissues.

Receptors and Signal Transduction Mechanisms
Cell signalling depends on receptors, which are proteins that interact with specific signalling molecules (ligands) to trigger a cellular response. Receptors can be categorized into two primary groups:

Membrane-Bound Receptors: These receptors reside on the cell membrane and engage with hydrophilic signalling molecules that are unable to traverse the lipid bilayer. When a ligand binds, receptors located in the membrane experience conformational alterations that trigger intracellular signalling pathways. Illustrations encompass:

G Protein-Coupled Receptors (GPCRs) represent a significant class of receptors that initiate intracellular signalling cascades via G proteins. They participate in a range of physiological processes, including vision, taste, and immune responses. Receptor Tyrosine Kinases (RTKs): Upon binding with their ligands, these receptors undergo dimerization and autophosphorylation, which leads to the activation of downstream signalling pathways. Receptor tyrosine kinases play a crucial role in the regulation of cellular growth, differentiation, and metabolic processes.

Intracellular receptors are situated within the cell, often found in the cytoplasm or nucleus. These molecules interact with hydrophobic signalling entities capable of traversing the cell membrane with ease, including steroid hormones like cortisol and oestrogen. Upon binding, these receptors have the capability to directly modulate gene expression through their role as transcription factors.

Essential Signalling Molecules: Signalling Molecules, Chemical Messengers, and Regulatory Proteins

Hormones are biochemical signals released by endocrine glands into the circulatory system. They oversee a multitude of physiological processes, encompassing metabolism, growth, and reproduction. Notable examples are insulin, adrenaline, and thyroid hormones.

Neurotransmitters are specialised chemicals that neurones release to convey signals across synapses to adjacent neurones or target cells. Notable examples are dopamine, serotonin, and acetylcholine, each of which is crucial for regulating mood, facilitating muscle contraction, and supporting cognitive functions.

Growth factors are proteins that promote the processes of cell proliferation, differentiation, and survival. They are essential for the processes of tissue development and repair. Instances encompass epidermal growth factor (EGF) and platelet-derived growth factor (PDGF).

Illustrations of Cellular Signalling Pathways

Cell signalling pathways consist of a sequence of intricate molecular events that culminate in a precise cellular response. In this discussion, we will delve into three critical signalling pathways: the MAPK/ERK pathway, the cAMP pathway, and the mechanisms of apoptosis signalling.

The MAPK/ERK signalling cascade
The MAPK/ERK pathway plays a vital role in the regulation of cellular processes such as growth, proliferation, differentiation, and survival.

Activation: The pathway initiates when growth factors interact with receptor tyrosine kinases (RTKs) located on the cell membrane. The interaction triggers receptor activation and sets off the process of autophosphorylation.

Signal transduction involves the recruitment of adapter proteins, such as Grb2, by activated receptor tyrosine kinases (RTKs), leading to the activation of the small GTPase Ras. Ras subsequently initiates a series of kinase activations, encompassing Raf, MEK, and ERK.

Upon activation, ERK moves to the nucleus, where it phosphorylates transcription factors, resulting in alterations in gene expression that facilitate cell division and differentiation.

The MAPK/ERK pathway plays a crucial role in numerous cellular processes and is frequently found to be dysregulated in cancers, underscoring its significance in cell signalling mechanisms.

The cyclic adenosine monophosphate (cAMP) signalling pathway
The cAMP (cyclic adenosine monophosphate) pathway serves as a crucial signalling mechanism employed by various hormones and neurotransmitters to influence target cells effectively.

Initiation: The process commences when a signalling molecule attaches to a G protein-coupled receptor (GPCR) located on the cell membrane, thereby activating a corresponding G protein.

Signal Transduction: The activated G protein triggers the enzyme adenylate cyclase, leading to the conversion of ATP into cyclic AMP (cAMP). cAMP functions as a crucial second messenger, enhancing the intracellular signal transduction.

The activation of protein kinase A (PKA) by cAMP triggers the phosphorylation of target proteins, resulting in a range of cellular responses, including enhanced glucose metabolism and alterations in gene expression.

The cAMP pathway plays a crucial role in the regulation of various physiological processes, such as glycogen breakdown, muscle contraction, and neurotransmitter release.

Signalling in Apoptosis
Apoptosis, often referred to as programmed cell death, plays an essential role in preserving tissue balance and removing damaged or unnecessary cells. Apoptotic signalling can be triggered via intrinsic or extrinsic pathways.

The extrinsic pathway is initiated by external signals, including death ligands like Fas ligand, which bind to death receptors located on the cell surface. This interaction triggers the activation of caspases, a group of proteolytic enzymes that carry out programmed cell death by cleaving designated cellular proteins.

Intrinsic Pathway: This pathway is initiated by internal signals, including DNA damage or cellular stress. In reaction to these signals, mitochondria discharge pro-apoptotic factors, including cytochrome c, which triggers the activation of caspases and results in apoptosis.

The process of apoptosis involves the convergence of both pathways on effector caspases, resulting in the orderly breakdown of the cell. This includes DNA fragmentation, membrane blebbing, and the creation of apoptotic bodies. Subsequently, neighbouring cells engage in phagocytosis of these entities, thereby inhibiting inflammation.

CHAPTER SEVEN

The Cell Cycle and Cell Division

The cell cycle comprises a sequence of stages that cells undergo to grow, duplicate their genetic material, and ultimately divide. This mechanism is essential for the growth, development, and repair of tissues in multicellular organisms. Effective control of the cell cycle is crucial for preserving the well-being of an organism. Dysregulation can result in significant repercussions, potentially culminating in cancer. This section will present a comprehensive overview of the cell cycle, elucidate the intricate processes of mitosis and meiosis, and investigate the connections between the cell cycle and cancer development.

Summary of the Cell Cycle
The cell cycle is composed of unique phases, each characterised by particular events that prime the cell for division. The cycle can be categorised into two primary stages: interphase and the mitotic phase (M phase).

Stages of the Cell Cycle
Interphase: This phase represents the most extended period of the cell cycle, characterised by cellular growth and preparation for division. Interphase is subdivided into three distinct stages:

During the G1 Phase, the cell undergoes growth while actively synthesising proteins and organelles. It also performs its standard metabolic activities. The G1 phase plays a vital role in determining cell size and energy reserves, essential prerequisites for DNA replication.
S Phase (Synthesis): In this phase, the process of DNA

replication takes place, leading to the duplication of chromosomes. At this stage, every chromosome is composed of two sister chromatids, poised for separation during the process of cell division.

G2 Phase (Gap 2): The cell undergoes further growth and readies itself for the process of mitosis. It produces proteins essential for mitosis and verifies that DNA replication has been executed accurately.

M Phase (Mitotic Phase): This phase encompasses the processes of mitosis and cytokinesis, during which the cell undergoes division to produce two daughter cells. Mitosis is categorised into specific stages: prophase, metaphase, anaphase, and telophase.

Regulatory Mechanisms and Checkpoints
The progression of the cell cycle is meticulously controlled by a series of checkpoints that oversee the advancement of the cell through each phase, guaranteeing that every stage is executed precisely before transitioning to the subsequent one. Essential checkpoints encompass:

The G1 Checkpoint, often referred to as the restriction point, evaluates the cell's size, energy reserves, and the integrity of its DNA. In unfavourable conditions, the cell can transition into a quiescent state (G0 phase) or initiate repair mechanisms for any incurred damage.

G2 Checkpoint: This checkpoint ensures the accurate replication of DNA and assesses any potential DNA damage. When problems arise, the cell cycle may be halted to facilitate necessary repairs.

The M Checkpoint, also known as the Spindle Checkpoint, is a critical regulatory mechanism that verifies the proper alignment and attachment of all chromosomes to the spindle apparatus prior to the transition into anaphase. This ensures that chromosomes are distributed evenly to the daughter cells.

Mitosis: A Form of Asexual Cell Division

The process of mitosis involves a single cell dividing to yield two genetically identical daughter cells, thereby preserving the chromosome number.

The Phases of Mitosis Prophase:

Chromatin undergoes condensation to form distinct chromosomes, with each chromosome comprising two sister chromatids that are connected at the centromere.
The nuclear envelope initiates its disassembly, while the mitotic spindle begins to assemble from the centrosomes, which migrate towards the opposing poles of the cell.
Metaphase:

The chromosomes position themselves along the metaphase plate, the equatorial plane of the cell, as a result of the pulling forces exerted by the spindle fibres that are anchored to the centromeres.
This alignment is essential for guaranteeing that each daughter cell obtains a precise and identical set of chromosomes.
Anaphase:

The sister chromatids are separated by the spindle fibres and migrate towards the opposing poles of the cell.
The centromeres undergo division, facilitating the separation of chromatids, which are then recognised as distinct chromosomes.
Telophase:

The chromosomes migrate to the poles and initiate the process

of decondensation, reverting to their chromatin form.
The nuclear envelope reassembles around each group of chromosomes, leading to the formation of two distinct nuclei within the cell.
Cytokinesis: Division of the Cytoplasmic Material
Cytokinesis occurs after mitosis and involves the division of the cytoplasm in a parent cell, resulting in the formation of two daughter cells. In animal cells, a contractile ring develops, constricting the cell membrane to produce two distinct cells. In plant cells, a cell plate emerges at the centre, ultimately maturing into a new cell wall that divides the two daughter cells.

Meiosis: A Process of Sexual Cell Division
Meiosis represents a distinct type of cell division that effectively halves the chromosome count, resulting in the formation of gametes (sperm and eggs) essential for sexual reproduction.

Function of Meiosis in Gamete Development
The main function of meiosis is to generate haploid gametes from diploid precursor cells. The decrease in chromosome number is crucial for guaranteeing that progeny obtain the appropriate quantity of genetic material when gametes combine during fertilisation.

Genetic Variation via Recombination and Independent Assortment
Meiosis generates genetic diversity via two fundamental mechanisms:

Recombination (Crossing Over): In prophase I of meiosis, homologous chromosomes align and engage in the exchange of segments of genetic material through a mechanism known as crossing over. This results in novel allele combinations on the chromosomes.

Independent Assortment: In metaphase I, homologous chromosomes align in a random manner at the metaphase plate. The random orientation leads to the independent assortment of chromosomes into gametes, thereby enhancing genetic diversity.

Meiosis involves two consecutive divisions: the first, known as meiosis I, serves as a reduction division, while the second, meiosis II, functions as an equational division, ultimately producing four distinct haploid cells.

The relationship between cancer and the cell cycle

Cancer encompasses a collection of diseases marked by unregulated cellular proliferation and growth, stemming from the disruption of the cell cycle mechanisms. This dysregulation can be linked to a range of genetic and environmental influences.

The Impact of Cell Cycle Dysregulation on Cancer Development

Cancer cells frequently evade the standard regulatory processes of the cell cycle, enabling them to multiply without restraint. Essential characteristics of cancer cells encompass:

Loss of checkpoint control can lead to a scenario where cancer cells bypass critical regulatory mechanisms of the cell cycle, enabling their progression despite adverse conditions such as DNA damage.

Enhanced Proliferative Signals: Tumour cells may generate their own proliferative signals or exhibit heightened sensitivity to growth factors, resulting in relentless activation of cellular division.

Inhibition of programmed cell death: Numerous cancer cells manage to bypass apoptosis, enabling them to persist for extended periods compared to typical cells.

Function of Oncogenes and Tumour Suppressors
Oncogenes: Oncogenes represent altered versions of standard genes (proto-oncogenes) that facilitate cellular proliferation and development. Activation of proto-oncogenes through mutations can result in uncontrolled cell growth, ultimately contributing to the development of cancer. Instances encompass the Ras gene and the Myc gene.

Tumour suppressor genes play a crucial role in regulating cell division and facilitating apoptosis. When these genes undergo mutations or are lost, their protective roles are diminished, leading to unchecked cell proliferation. A prominent tumour suppressor gene is p53, commonly called the "guardian of the genome" due to its function in inhibiting the division of damaged cells.

CHAPTER EIGHT

Cellular Transport Mechanisms

Transport mechanisms within cells play a crucial role in sustaining homeostasis and enabling interaction with the surrounding environment. These mechanisms can be categorised into two primary types: passive transport and active transport. Every transport mechanism is essential in governing the movement of substances through the cell membrane, guaranteeing the proper functioning of the cell.

Facilitated Movement
Passive transport refers to the movement of molecules across the cell membrane that occurs without the use of energy (ATP). This mechanism utilises the inherent kinetic energy of molecules, facilitating their movement along the concentration gradient, transitioning from regions of higher concentration to those of lower concentration. Passive transport encompasses essential processes such as diffusion, osmosis, and facilitated diffusion.

Diffusion
Diffusion refers to the movement of molecules from regions of higher concentration to those of lower concentration, continuing until a state of equilibrium is achieved. This occurrence arises from the stochastic motion of molecules, leading to their natural dispersion within a solution or across a membrane.

Essential Features of Diffusion:
Molecules will move along their concentration gradient until

equilibrium is reached, resulting in no net movement.
Categories of Molecules: Small nonpolar molecules, like oxygen and carbon dioxide, can readily pass through the lipid bilayer of the cell membrane via diffusion.
The rate of diffusion is influenced by various factors, such as molecular size, temperature, and the concentration gradient. For instance, elevated temperatures enhance molecular activity, accelerating the process of diffusion.

Osmosis

Osmosis is a distinct form of diffusion that involves the movement of water molecules through a semipermeable membrane. Water transitions from regions of reduced solute concentration to those with elevated solute concentration, striving to balance solute levels across the membrane.

Essential Features of Osmosis:

Aquaporins: Specialised protein channels known as aquaporins enable the swift passage of water molecules through the cell membrane, facilitating rapid water movement.
Tonicity: The movement of water influences the cell's volume and shape, contingent upon the tonicity of the surrounding

solution:
An isotonic solution occurs when the solute concentrations are balanced both inside and outside the cell, leading to no net movement of water across the membrane.
A hypotonic solution is characterised by a lower solute concentration compared to the inside of the cell, resulting in water influx that may cause the cell to swell or even undergo lysis.
In a hypertonic solution, the external environment possesses a greater concentration of solutes compared to the intracellular

fluid. This disparity prompts the efflux of water from the cell, potentially resulting in cell shrinkage or crenation.

Facilitated diffusion

Facilitated diffusion is a mechanism that enables larger or polar molecules to traverse the cell membrane through the assistance of specialised transport proteins. This approach operates without the need for energy, as molecules continue to traverse their concentration gradient.

Essential Features of Facilitated Diffusion:

Transport Proteins: Facilitated diffusion employs two distinct classes of proteins:

Channel proteins create openings that enable the selective passage of certain molecules. For instance, ion channels facilitate the transport of ions such as sodium, potassium, and calcium.

Carrier proteins interact with specific molecules, undergoing conformational changes to facilitate their movement across the membrane. For instance, glucose transporters enable the absorption of glucose into cellular structures.

Saturation occurs when the rate of facilitated diffusion reaches its peak, as all available transport proteins become occupied, resulting in a state of saturation.

Elements Influencing Passive Transport

Various elements affect the effectiveness and speed of passive transport processes:

A steeper gradient increases diffusion rates due to a more significant difference in concentration between two regions.

Temperature: Elevated temperatures boost molecular activity, facilitating diffusion and osmosis.

Smaller molecules diffuse more rapidly than their larger counterparts, as they encounter less resistance while traversing the membrane.

The lipid composition and structural characteristics of the cell membrane significantly influence the permeability of substances, determining their ease of passage through this barrier.

Active transport
Active transport entails the translocation of molecules contrary to their concentration gradient, necessitating energy in the form of ATP. This process is essential for regulating the levels of ions and various substances within the cell, ensuring they remain distinct from those found in the external environment.

Sodium-Potassium Pump
The sodium-potassium pump (Na+/K+ pump) serves as a classic illustration of active transport mechanisms. This process is crucial for sustaining the electrochemical gradient across the cell membrane, a fundamental requirement for the transmission of nerve impulses and the contraction of muscles.

Essential Characteristics of the Sodium-Potassium Pump:
The pump facilitates the active transport of sodium ions from the cell while simultaneously bringing potassium ions into the cell, generally maintaining a ratio of 3 sodium ions expelled for every 2 potassium ions imported.
Energy Requirement: This mechanism necessitates ATP to alter the conformation of the pump, enabling the transport of ions against their concentration gradients.
Significance: The pump plays a vital role in sustaining the resting membrane potential and is essential for numerous cellular processes, such as nutrient absorption and waste elimination.
Endocytosis and Exocytosis
Alongside pumps and transport proteins, cells employ bulk

transport mechanisms to facilitate the movement of larger molecules and particles across the membrane.

Endocytosis: This process entails the cell membrane enveloping substances, subsequently folding inward to create vesicles that transport materials into the cell. Various forms of endocytosis exist:

Phagocytosis, commonly known as "cell eating," is a vital process where cells engulf substantial particles, including bacteria or cellular debris. This results in the formation of a phagosome, which subsequently fuses with lysosomes to facilitate digestion.

Pinocytosis, often referred to as "cell drinking," is the process by which cells internalise small droplets of extracellular fluid along with dissolved solutes. The membrane undergoes invagination, resulting in the formation of vesicles that facilitate the transport of these fluids into the cell.

Exocytosis refers to the mechanism through which cells release substances that are enclosed within vesicles. This mechanism plays a vital role in the secretion of hormones, neurotransmitters, and various other substances. In the process of exocytosis:

Vesicles that harbour the material merge with the plasma membrane, discharging their contents into the extracellular environment.

This mechanism is crucial for intercellular communication and the elimination of metabolic byproducts.

Mass Movement

Mechanisms of bulk transport, including endocytosis and exocytosis, are essential for numerous cellular functions, such as:

Nutrient Uptake: Endocytosis enables cells to internalise large molecules or particles that are unable to traverse the membrane through passive or facilitated transport mechanisms. Cell Communication: The process of exocytosis plays a crucial role in the release of signalling molecules, facilitating intercellular communication.

Immune Response: The process of phagocytosis plays a vital role in the immune system, enabling immune cells to capture and eliminate pathogens effectively.

Cellular Transport and Its Role in Homeostasis

Transport mechanisms within cells are essential for sustaining homeostasis both at the cellular level and throughout the entire organism. Homeostasis denotes the equilibrium within the internal environment that is essential for cells to operate at their best. Here are several important mechanisms through which cellular transport plays a vital role in maintaining homeostasis.

Active transport mechanisms, including the sodium-potassium pump, play a crucial role in maintaining the concentrations of vital ions and nutrients within cells. The regulation of ion concentrations is essential for fundamental biological processes such as nerve signalling and muscle contraction.

Cells are required to effectively dispose of waste products produced during metabolic processes. Exocytosis plays a crucial role in the expulsion of waste materials, thereby contributing to the maintenance of a stable internal environment.

The regulation of fluid balance is crucial, as osmosis significantly influences the movement of water across cellular membranes, thereby ensuring the maintenance of cellular volume and pressure. This is essential for mechanisms such as nutrient uptake and waste elimination.

Ion transport across the membrane plays a critical role in modulating the pH levels of both the cytoplasm and the extracellular fluid. The transport of bicarbonate ions plays a crucial role in maintaining blood pH homeostasis.

Response to Environmental Changes: The mechanisms of cellular transport enable cells to adjust to fluctuations in their surroundings, including shifts in nutrient supply, temperature, and osmotic pressure. This adaptability is essential for thriving in ever-changing environments.

CHAPTER NINE

Genetic Inheritance and Molecular Genetics

The transmission of genetic material involves the transfer of genes, which are made up of DNA, from one generation to the next, from parents to their progeny. The transfer of genetic information plays a crucial role in defining the traits and characteristics of living organisms. The exploration of genetics includes the foundational principles of inheritance laid out by Gregor Mendel, as well as the intricate molecular processes that govern gene expression and regulation. Grasping the intricacies of inheritance, mutations, and gene regulation is essential for comprehending the vast complexity of biological diversity and the mechanisms underlying various disorders.

Mendelian principles and the mechanisms of inheritance
Gregor Mendel, a pioneer in the field of genetics, laid down essential principles of inheritance by conducting meticulous experiments with pea plants. Mendel's research laid the foundational principles for the field of genetics, elucidating the mechanisms by which traits are passed down through successive generations.

The principle of segregation states that every individual possesses two alleles for a specific trait, inherited from each parent. During the formation of gametes, these alleles separate, ensuring that each gamete contains only a single allele for each trait.

The principle of Independent Assortment states that alleles for various traits segregate independently during the formation of gametes, facilitating the emergence of novel trait combinations.

These principles elucidate the mechanisms by which traits manifest in consistent ratios among progeny, assuming the traits adhere to Mendelian inheritance frameworks. A dominant allele will obscure the expression of a recessive allele when both are present, yet recessive traits may resurface in future generations.

Dominant and Recessive Characteristics
In the context of dominant and recessive inheritance, traits are governed by pairs of alleles. A dominant allele necessitates just one copy for the trait to manifest, while a recessive allele demands two copies for expression.

For instance, when "A" denotes a dominant allele associated with brown eyes and "a" signifies a recessive allele linked to blue eyes, individuals possessing either AA or Aa genotypes will exhibit brown eyes, whereas only those with the aa genotype will display blue eyes. The arrangement of these characteristics adheres to discernible patterns that can be illustrated using Punnett squares.

Punnett Squares and Genetic Probability
A Punnett square serves as a visual tool to forecast the genetic likelihood of offspring by analysing the genotypes of the parental organisms. Organising the potential alleles from each parent facilitates the calculation of probabilities for various genotypic and phenotypic outcomes.

For instance, when both parents possess a heterozygous genotype for a specific trait (Aa), the Punnett square forecasts a ratio of 1:2:1 for the genotypes AA, Aa, and aa in their progeny. This ratio results in a 3:1 phenotypic ratio when A is considered dominant and an is regarded as recessive.

Alterations in DNA and Their Impact on Genetic Conditions
Mutations represent alterations in the DNA sequence, leading to a range of effects on gene functionality. Although numerous mutations may be neutral, certain ones can lead to genetic disorders by modifying the structure or expression of proteins.

Categories of Genetic Alterations
Point mutations involve the alteration of a single nucleotide, potentially leading to various outcomes: a missense mutation, where a different amino acid is incorporated; a nonsense mutation, which introduces a premature stop codon; or a silent mutation, which leaves the protein unchanged.

Insertions involve the addition of one or more nucleotides to the DNA sequence, potentially disrupting the gene's reading frame, a phenomenon referred to as a frameshift mutation.

Deletions involve the removal of one or more nucleotides, potentially causing a frameshift that can lead to substantial alterations in the protein structure.

Understanding the Mechanisms Behind Genetic Disorders Induced by Mutations
Mutations in essential genes can result in genetic disorders. A specific point mutation in the haemoglobin gene leads to sickle cell anaemia, resulting in the production of an altered protein that causes red blood cells to adopt a misshapen form. Similarly, alterations such as insertions or deletions in crucial regulatory genes can result in conditions like cystic fibrosis, where the loss of three nucleotides impacts a chloride ion transporter protein,

causing significant respiratory and digestive complications.

Regulation of Gene Expression
Gene regulation guarantees that the appropriate genes are activated in the correct cells at the precise moment. This regulation is crucial for ensuring appropriate development and optimal cellular function.

Operons in prokaryotic organisms
In prokaryotic organisms, the regulation of gene expression frequently occurs through operons, which are groups of genes that are transcribed simultaneously into a single mRNA molecule. The lac operon in E. coli is the most recognised operon, playing a crucial role in the regulation of lactose metabolism. The operon is regulated by a repressor protein that attaches to the operator region, thereby inhibiting transcription. In the presence of lactose, it interacts with the repressor, facilitating the transcription of genes essential for the metabolism of lactose by RNA polymerase.

Regulation of Gene Expression and Epigenetic Mechanisms in Eukaryotic Systems
In eukaryotic organisms, the regulation of genes exhibits a higher level of complexity, incorporating various layers of control. This includes epigenetics, which pertains to modifications in gene expression that do not involve changes to the DNA sequence itself. A significant process involves the addition of methyl groups to cytosine residues in DNA, which generally leads to the silencing of gene expression. Another aspect is histone modification, where chemical alterations to the proteins that DNA wraps around affect the accessibility of genes for transcription.

Epigenetic regulation enables cells to adapt to environmental fluctuations and can be passed down, influencing not only the individual but also potentially affecting future generations.

Genetic manipulation and biotechnological advancements
Recent developments in biotechnology have allowed for the unprecedented manipulation of genes. Genetic engineering entails the alteration of an organism's genome through technological means, resulting in significant advancements in medicine, agriculture, and various other domains.

CRISPR and Gene Modification
CRISPR-Cas9 represents a groundbreaking advancement in gene editing, enabling researchers to accurately target and cleave DNA at designated sites. This system, derived from a bacterial defence mechanism, employs a guide RNA to precisely direct the Cas9 enzyme to a specific DNA sequence, facilitating a targeted cut. The cell subsequently initiates repair mechanisms for the break, facilitating the insertion, deletion, or modification of genetic material.

CRISPR presents promising avenues for addressing genetic disorders, including muscular dystrophy and cystic fibrosis, through the correction of the fundamental mutations involved. This approach also offers potential for developing crops that resist diseases and for the engineering of organisms aimed at environmental applications.

Cloning and Genetically Modified Organisms
Cloning entails the process of producing a genetically identical replica of an organism. Dolly the sheep stands out as a landmark case, being the first mammal successfully cloned from an adult somatic cell. Cloning serves vital roles in the conservation of endangered species, alongside its significance in research and

agricultural advancements.

Transgenic organisms are genetically altered entities that incorporate genes from different species. These organisms are employed in agricultural practices to develop crops that exhibit resistance to pests, diseases, or herbicides. For example, Bt maize has been modified to generate a toxin derived from the bacterium Bacillus thuringiensis, which targets specific insects while remaining safe for human consumption.

In the field of medicine, transgenic animals serve a crucial role in the production of pharmaceuticals. For instance, insulin can be derived from genetically modified bacteria, while antithrombin, a protein essential for preventing blood clots, is produced using genetically modified goats.

CHAPTER TEN

Introduction to Cell Differentiation and Development

The processes of cell differentiation and development are fundamental, guiding the transformation of a single fertilised egg into a remarkably intricate multicellular organism. Every multicellular organism, irrespective of its size or complexity, originates from a singular cell that undergoes division, specialisation, and organisation to form a body composed of various tissues, organs, and systems. Exploring the process by which a single cell transforms into a fully functional organism requires an in-depth examination of cellular differentiation, gene regulation, tissue formation, and organ development.

From a Singular Cell to a Multifaceted Organism
Development initiates with a zygote, the singular cell created upon the fertilisation of an egg by a sperm. The zygote experiences a swift process of cell division, referred to as cleavage, resulting in the formation of a cellular mass. The initial cells, known as blastomeres, begin as identical entities but quickly start to differentiate in their roles. During differentiation, cells start to specialise, leading to the formation of the various cell types present in a mature organism.

The process of differentiation is crucial for development, as it facilitates the creation of various tissues and organs, each serving distinct functions. The progression of an organism is intricately controlled by a blend of genetic elements and signalling mechanisms that dictate the destiny of each cell.

Stem Cells and Pluripotency

Central to the process of differentiation are stem cells, which are undifferentiated entities possessing the extraordinary capacity to proliferate and generate diverse cell types. Stem cells can be categorised according to their ability to differentiate:

Totipotent stem cells represent the pinnacle of cellular versatility, possessing the remarkable ability to differentiate into any cell type, including the placenta. The zygote and the cells generated during the initial divisions exhibit totipotency.

Pluripotent stem cells: These cells, located in the inner cell mass of the blastocyst during early development, possess the remarkable ability to differentiate into almost any cell type within the body, with the exception of the placenta. These are the progenitor cells utilised in various domains of regenerative therapies.

Multipotent stem cells: These cells exhibit a higher degree of specialisation and possess the capability to differentiate into a restricted variety of cell types. For instance, haematopoietic stem cells located in the bone marrow have the capacity to differentiate into various blood cell types, yet they lack the ability to transform into other tissue types such as muscle or nerve cells.

The ability of embryonic stem cells to become any type of cell is essential for the initial stages of development, as they form all the distinct tissues and organs within the organism.

The process of differentiation and its significance in developmental biology

Cell differentiation refers to the transformation of unspecialised cells into specialised cells that perform unique functions. The process of differentiation is driven by the activation or repression of particular genes, influenced by a range of internal and external signals. This process is essential for the differentiation of various cell types, including neurones, muscle cells, and epithelial cells.

The destiny of a cell is influenced by its location within the developing embryo and the cues it obtains from adjacent cells. The signals orchestrate the expression of particular genes, resulting in the synthesis of proteins that determine the architecture and role of the specialised cell.

The process of differentiation is generally irreversible for the majority of cell types. Once a cell has specialised, it usually cannot revert to an undifferentiated state. However, there are notable exceptions, such as induced pluripotent stem cells (iPSCs), which allow differentiated cells to be reprogrammed back into a pluripotent state.

Processes of Growth and Differentiation
The intricate process of development is meticulously governed by gene regulation and complex signalling pathways, which guarantee that cells differentiate correctly and assemble into functional tissues and organs.

Regulation of Genes During Development
The regulation of gene expression during developmental processes involves the intricate interplay of transcription factors and signalling molecules. Transcription factors are proteins that interact with DNA, playing a crucial role in either facilitating or suppressing gene transcription, thereby regulating the synthesis of proteins vital for differentiation.

Throughout the process of development, various transcription factors are selectively expressed in distinct regions of the embryo, facilitating the specialisation of cells into their respective types in each area. For instance, homeobox (Hox) genes represent a category of transcription factors that are essential in establishing the body plan of an organism, directing the development of structures in their appropriate locations along the anterior-posterior axis (from head to tail).

Stages of Embryonic Development
Development advances through clearly defined phases:

Fertilisation involves the fusion of sperm and egg, resulting in the formation of a zygote, which possesses totipotency and signifies the initiation of developmental processes.

Cleavage: The zygote experiences a series of rapid cell divisions, resulting in a compact cluster of smaller cells without an increase in overall size.

During blastulation, the cells organise to create a hollow structure known as a blastocyst. The inner cell mass is responsible for developing into the embryo, whereas the outer cells contribute to the formation of the placenta.

During gastrulation, the blastocyst's cells undergo a remarkable rearrangement, forming three distinct layers known as the germ layers: ectoderm, mesoderm, and endoderm. These layers serve as the foundational elements for the development of all tissues and organs.

Ectoderm: Gives rise to the skin, brain, and nervous system. The mesoderm gives rise to the muscles, bones, blood vessels,

and heart.
The endoderm gives rise to the epithelial lining of the gastrointestinal tract, respiratory system, and various internal organs.
Organogenesis involves the differentiation of the three germ layers into distinct tissues and organs. For example, the ectoderm gives rise to the nervous system, whereas the mesoderm differentiates into muscles and connective tissues.

Development and Differentiation of Tissues
Upon differentiation, cells initiate the formation of the diverse tissues that constitute the organism. The four fundamental types of tissues are as follows:

Epithelial tissue serves as the protective outer layer of the body and lines the internal organs. It performs protective, secretory, and absorptive roles. Examples of epithelial tissues include the skin and the lining of the digestive tract.

Connective tissue encompasses a variety of tissues, including bone, blood, and cartilage. Connective tissue serves as a fundamental framework, offering structural integrity and facilitating the interconnection of various tissues. Blood serves as a vital medium for the transport of nutrients and oxygen, whereas bones and cartilage play crucial roles in providing structural integrity and facilitating movement.

Muscle tissue is uniquely adapted for contraction, facilitating movement throughout the organism. Muscle tissue can be categorised into three distinct types:

Skeletal muscle: Muscles under voluntary control that facilitate the movement of bones.
Smooth muscle refers to the involuntary muscle tissue located

in various organs, such as the stomach and blood vessels.
Cardiac muscle is the involuntary muscle found within the heart.
Nervous tissue consists of neurones and glial cells, playing a crucial role in signal transmission across the body, facilitating communication among various bodily systems. It is essential for both voluntary and involuntary reactions to stimuli.

Development of Organ Systems
Throughout development, tissues intricately arrange themselves into functional organ systems that collaborate to uphold homeostasis and execute vital life processes. Organ systems encompass:

The nervous system, originating from the ectoderm, differentiates into the brain, spinal cord, and peripheral nerves, orchestrating bodily functions and adapting to environmental stimuli.

The circulatory system comprises the heart and blood vessels, which originate from the mesoderm. This system is responsible for transporting oxygen, nutrients, and hormones to cells while efficiently removing waste products such as carbon dioxide.

The respiratory system comprises the lungs and airways, originating from the endoderm, and plays a crucial role in gas exchange by delivering oxygen to the bloodstream and removing carbon dioxide.

The digestive system, originating from the endoderm, is responsible for processing food, absorbing essential nutrients, and expelling waste products.

The musculoskeletal system comprises bones and muscles,

both derived from the mesoderm, which serve to provide structural integrity and facilitate movement.

The reproductive system comprises organs that originate from the mesoderm, playing a crucial role in species propagation by facilitating gamete production and supporting offspring development.

CHAPTER ELEVEN

Introduction to Immunology: Cells and Molecules in the Immune Response

The immune system serves as the body's protective barrier against various pathogens, including viruses, bacteria, fungi, and parasites. This intricate system comprises various cells, tissues, and molecules that collaborate to identify and eradicate foreign entities while differentiating them from the body's own cellular components. The exploration of immunology focusses on the immune system's intricate components and their interactions, orchestrating a unified response to safeguard the body against disease. The immune response is categorised into two primary components: innate immunity and adaptive immunity.

Innate and Adaptive Immunity
The immune system functions through two distinct mechanisms: innate immunity, which is nonspecific, and adaptive immunity, which is specific.

The innate immune system serves as the initial barrier against pathogens in the body. The system comprises physical barriers, including the skin and mucous membranes, alongside various cells and molecules that deliver a swift yet nonspecific reaction to intruders. Innate immunity encompasses various immune cells, including macrophages, dendritic cells, and natural killer (NK) cells. These cells recognise pathogens via broad molecular signatures known as pathogen-associated molecular patterns

(PAMPs), which are prevalent among various microbes. The innate immune system responds rapidly; however, it lacks the capacity for enduring defence against particular pathogens.

The response of adaptive immunity is characterised by a slower onset, yet it exhibits a remarkable specificity. This process engages distinct cell types, including T cells and B cells, which identify particular antigens—unique molecules present on the surface of pathogens. The adaptive immune system possesses a remarkable capacity to retain information about pathogens following the first encounter, facilitating a more rapid and robust response during future exposures. This phenomenon of immunological memory underpins the principle of vaccination.

Cellular Participants in the Immune Response
The immune response is orchestrated by a variety of cell types, each fulfilling specific functions in the defence of the organism:

Macrophages: These sizable phagocytic cells play a crucial role in the innate immune system. They internalise and break down pathogens and necrotic cells. Macrophages are essential in signalling the adaptive immune system through the presentation of antigens (foreign molecules) to T cells, effectively connecting the innate and adaptive responses.

Dendritic Cells: These cells play a vital role as antigen-presenting cells (APCs). They engage with pathogens, process them, and present fragments of these pathogens (antigens) on their surface for detection by T cells, thereby triggering the adaptive immune response.

T Cells: These are pivotal components of the adaptive immune response. There are various types of T cells:

Helper T cells (CD4+ T cells) play a crucial role in coordinating the immune response by activating various immune cells, such as B cells and cytotoxic T cells, through the secretion of signalling molecules known as cytokines.

Cytotoxic T cells (CD8+ T cells) are specialised immune cells that identify and eliminate infected cells, especially those compromised by viral infections.

B cells play a crucial role in the immune response by generating antibodies that effectively neutralise pathogens. When an antigen is encountered, B cells have the ability to differentiate into plasma cells that produce significant quantities of antibodies tailored to the specific pathogen. B cells are also involved in the maintenance of immunological memory.

Natural Killer (NK) Cells: These cells play a crucial role in the innate immune system, effectively targeting and eliminating virus-infected and cancerous cells without requiring prior activation by antigens.

Immunoglobulins and Immunogens

Antibodies, known as immunoglobulins, are Y-shaped proteins synthesised by B cells. They play a vital role in recognising and eliminating pathogens such as bacteria and viruses. Antibodies interact with antigens, which are foreign entities, usually proteins or polysaccharides, located on the surfaces of pathogens.

Architecture and Role of Immunoglobulins

An antibody molecule is composed of four polypeptide chains: two heavy chains and two light chains, structured in a Y configuration. The tips of the Y structure constitute the variable regions, exhibiting a high degree of specificity for distinct antigens. These areas facilitate the binding of the antibody to a specific antigen with great affinity, thereby enabling the immune system to effectively target particular pathogens.

The constant region of the antibody plays a crucial role in defining its class and function within the immune response. **There exist five primary categories of antibodies:**

IgG is the predominant antibody found in the bloodstream, playing a crucial role in sustaining long-term immunity.
IgA: Present in mucosal surfaces, including the respiratory and gastrointestinal systems.
IgM is the initial antibody generated in response to an immune challenge.
IgE: Engaged in the mechanisms of allergic responses.
IgD: Present on the surface of B cells, it is crucial for the initiation of B cell activation.
Antibodies play a crucial role in neutralising pathogens through various mechanisms:

Inhibiting the pathogen's entry into cells (neutralisation).
Labelling the pathogen for elimination by phagocytic cells (opsonisation).
Initiating the complement system, resulting in the elimination of the pathogen.
Mechanisms of Cellular Recognition of Exogenous Molecules
The immune system's cells, especially T cells and B cells, identify pathogens via their antigen receptors. Every T or B cell possesses a distinct receptor that interacts with a particular antigen. The variety of receptors enables the immune system to identify a wide range of pathogens.

T cell receptors (TCRs) identify antigens displayed on the surfaces of infected cells or antigen-presenting cells (APCs), including macrophages and dendritic cells, alongside major histocompatibility complex (MHC) molecules.

B cell receptors (BCRs) identify antigens in their natural state, resulting in the activation of B cells and the subsequent production of antibodies.

Vaccination and Immune Memory

Vaccination serves as an effective strategy for the prevention of infectious diseases, utilising the body's inherent immune mechanisms to create immunological memory while avoiding the onset of illness.

The Mechanism of Vaccine Functionality

Vaccines function by delivering a non-harmful variant of a pathogen or its constituents (like proteins or inactivated viruses) into the organism. This initiates the immune system's ability to identify the foreign antigen and activate a response.

Following vaccination, the activation of B cells and T cells occurs, resulting in the generation of antibodies and the establishment of memory cells. These memory cells endure within the organism for extended durations, frequently throughout the lifespan.

Upon exposure to the actual pathogen, a vaccinated individual's immune system exhibits a rapid and efficient response, frequently thwarting infection or reducing the severity of symptoms. This phenomenon is referred to as protective immunity.

Function of Memory Cells in Immune Response

Memory cells represent a defining feature of the adaptive immune system, serving a vital function in sustaining long-term immunity. Following an infection or vaccination, the immune system generates both memory B cells and memory T cells.

Memory B cells retain the information of the specific antigen, allowing for a rapid antibody response upon re-exposure to the pathogen. The cells maintain a quiescent state until reactivation occurs in response to the same antigen's presence.

Memory T cells, encompassing both helper and cytotoxic types, exhibit remarkable longevity and can swiftly initiate activation upon subsequent encounters with the pathogen, resulting in a prompt and robust immune response.

The existence of memory cells guarantees a more rapid response from the immune system during later encounters with the same pathogen, often neutralising it before an infection can establish itself. This principle is fundamental to the effectiveness of vaccination initiatives in managing diseases such as measles, polio, and influenza.

CHAPTER TWELVE

Advances in Molecular and Cell Biology

Recent decades have seen remarkable progress in the fields of molecular and cell biology, primarily fuelled by technological innovations that have transformed our comprehension of life at the molecular scale. These advancements have not only illuminated essential biological mechanisms but also paved the way for innovative approaches to disease treatment, genetic manipulation, and the promotion of regenerative therapies. In this exploration, we will investigate the significant breakthroughs in the study of biological molecules and cellular processes, highlighting innovative technologies, stem cell therapies, CRISPR gene editing, and the emerging trends that are influencing this ever-evolving discipline.

Innovative Approaches and Methodologies
The emergence of advanced technologies such as Polymerase Chain Reaction (PCR), DNA sequencing, and improvements in microscopy has revolutionised the field, enabling researchers to manipulate, visualise, and comprehend genetic material and cellular mechanisms with unparalleled precision.

Polymerase Chain Reaction (PCR)
PCR is an innovative method introduced in the 1980s that enables the amplification of targeted DNA sequences. Through the use of PCR, researchers are able to generate millions or even billions of copies of a specific DNA fragment, establishing it as an essential technique in the field. Its applications span

various fields, including medical diagnostics for detecting viral infections such as HIV or COVID-19, as well as forensic science and evolutionary biology. The rapidity, precision, and capacity to utilise minuscule quantities of DNA render this technique an essential asset for both investigative and clinical purposes.

Recent innovations, including real-time PCR (qPCR), enable the precise quantification of DNA as it occurs, proving essential for diagnostic applications, particularly in monitoring the advancement of diseases or infections. The polymerase chain reaction has been instrumental in facilitating swift and precise genetic testing, thereby advancing the field of personalised medicine.

DNA Sequencing
The process of DNA sequencing involves elucidating the exact arrangement of nucleotides in a DNA molecule. The initial comprehensive sequence of the human genome was released in 2003, after more than ten years of extensive research efforts. Today, advancements in sequencing technologies allow for the rapid sequencing of entire genomes in mere days instead of years, and at a significantly reduced cost. This has transformed disciplines including genomics, evolutionary studies, and cancer investigation.

Next-generation sequencing technologies have enabled researchers to pinpoint genetic mutations linked to various diseases, laying the groundwork for precision medicine, which customises treatments to the unique genetic profiles of individual patients. This holds significant importance in the field of oncology, as sequencing cancer genomes can uncover mutations that promote tumour growth, paving the way for tailored treatment approaches.

Innovations in Microscopy

Microscopy has historically served as a fundamental instrument in the study of cells, allowing scientists to observe cellular structures and their intricate details. Recent advancements in microscopy techniques, including super-resolution microscopy and live-cell imaging, have expanded the limits of observable phenomena.

Advanced microscopy techniques, including STORM (Stochastic Optical Reconstruction Microscopy) and PALM (Photoactivated Localisation Microscopy), enable researchers to observe structures at an incredibly detailed level, surpassing the limitations of conventional light microscopy. This advancement allows for the observation of the intricate organisation of proteins, lipids, and various other molecules within cells with remarkable precision.

Techniques for live-cell imaging, combined with fluorescent markers, enable the real-time observation of cellular processes. This has yielded essential understanding of dynamic mechanisms such as cell division, intracellular transport, and protein trafficking.

Stem Cells and Regenerative Medicine

One of the most promising areas of research is the field of stem cells and regenerative medicine. Stem cells represent a unique class of undifferentiated cells, possessing the remarkable ability to differentiate into various cell types within the organism. They are essential for development, tissue upkeep, and healing processes. The identification of pluripotent stem cells—cells that can differentiate into any cell type within the organism—has transformed the fields of medicine and biology.

Possibilities for Disease Treatment

The exploration of stem cells presents significant opportunities for addressing conditions characterised by the depletion or impairment of particular cell populations. Embryonic stem cells and induced pluripotent stem cells possess the remarkable ability to differentiate into any cell type, presenting a promising avenue for the regeneration of damaged tissues or organs.

Several possible applications are as follows:

Parkinson's disease: Researchers are investigating methods to generate dopamine-producing neurones from stem cells, aiming to replace the neurones that are lost in this condition.
Diabetes: The potential of stem cells to differentiate into insulin-producing pancreatic cells offers a promising avenue for developing a cure for type 1 diabetes.
Heart disease: The application of stem cells holds potential for the regeneration of damaged cardiac tissue after a myocardial infarction.
The field of regenerative medicine encompasses the advancement of 3D bioprinting methods, utilising stem cells to cultivate tissues or potentially entire organs in laboratory settings. This may offer a viable solution to the challenges associated with organ transplants, particularly the scarcity of available donor organs.

Even with these progressions, obstacles persist, such as the possibility of tumour development from stem cells and the risk of immune rejection. Nonetheless, investigations continue, and the outlook for stem cell treatments appears optimistic.

CRISPR and Genetic Engineering
CRISPR (Clustered Regularly Interspaced Short Palindromic Repeats) has emerged as a highly effective tool in the field of biology, enabling accurate genetic alterations. The CRISPR-Cas9

system, originating from bacterial immune mechanisms, enables researchers to precisely "cut" DNA at targeted sites and implement genetic modifications.

Ethical Implications in Genetic Modification

CRISPR presents remarkable possibilities for addressing genetic disorders like cystic fibrosis, sickle cell anaemia, and certain cancers, yet it simultaneously brings forth considerable ethical dilemmas. A significant concern revolves around germline editing, as modifications to the genome may be inherited by subsequent generations. This has ignited discussions regarding the ethical ramifications of genetically engineered offspring and the potential for unforeseen outcomes.

Moreover, there are apprehensions regarding the off-target effects associated with CRISPR, where unintended alterations to the genome may take place. With advancements in technology, the associated risks are being reduced; however, ethical considerations continue to play a pivotal role in the ongoing discussion regarding the appropriate timing and application of gene editing in humans.

Emerging Directions in Cellular and Molecular Research

The future of this scientific field is filled with exciting possibilities. As advancements in technology progress, numerous significant trends are influencing the trajectory of research and its possible applications.

Tailored Therapeutics

Recent progress in genomics, along with innovative technologies such as CRISPR and next-generation sequencing, is propelling the movement towards tailored medical approaches. By comprehending a person's genetic composition, one can customise therapies to meet their unique requirements. This holds significant importance in cancer treatment, as the specific genetic mutations present in an individual's tumour can inform and direct therapeutic strategies.

Organoids and the Engineering of Tissues
Organoids represent small, simplified replicas of organs that are cultivated in vitro using stem cells. This offers a robust framework for investigating human development, disease mechanisms, and pharmaceutical evaluation. Organoids are being engineered for multiple organs, such as the brain, liver, and kidney, enabling researchers to investigate diseases in unprecedented ways.

The field of tissue engineering is rapidly advancing, focussing on the cultivation of functional tissues in laboratory settings, with the potential for future applications in transplantation. Progress in biomaterials and stem cell research is leading to the creation of functional tissues, including skin, cartilage, and potentially whole organs.

Synthetic Biology
Synthetic biology represents a convergence of disciplines focused on the innovative redesign of organisms, enabling them to acquire novel functionalities for practical applications. This may encompass the engineering of microorganisms capable of generating biofuels, breaking down environmental contaminants, or even synthesising novel materials.

The fusion of artificial intelligence with synthetic biology is anticipated to enhance discoveries by enabling researchers to simulate intricate biological systems and forecast the results of genetic alterations with greater precision.

Genomic Analysis at the Single-Cell Level

Recent progress in single-cell genomics enables scientists to investigate gene expression at the individual cell level. This has yielded essential insights into the diversity present within tissues and tumours, enhancing our comprehension of cancer, neurodegenerative disorders, and developmental processes.

THE END